Zoo
Julie Murray

Abdo Kids Junior
is an Imprint of Abdo Kids
abdobooks.com

Abdo
FIELD TRIPS
Kids

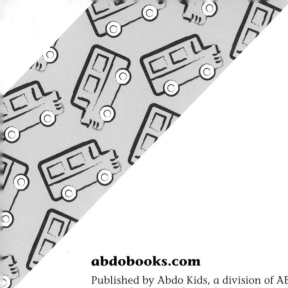

abdobooks.com

Published by Abdo Kids, a division of ABDO, P.O. Box 398166, Minneapolis, Minnesota 55439. Copyright © 2020 by Abdo Consulting Group, Inc. International copyrights reserved in all countries. No part of this book may be reproduced in any form without written permission from the publisher. Abdo Kids Junior™ is a trademark and logo of Abdo Kids.

Printed in the United States of America, North Mankato, Minnesota.

102019
012020

Photo Credits: Alamy, iStock, Shutterstock

Production Contributors: Teddy Borth, Jennie Forsberg, Grace Hansen

Design Contributors: Christina Doffing, Candice Keimig, Dorothy Toth

Library of Congress Control Number: 2019941212
Publisher's Cataloging-in-Publication Data

Names: Murray, Julie, author.
Title: Zoo / by Julie Murray
Description: Minneapolis, Minnesota : Abdo Kids, 2020 | Series: Field trips | Includes online resources and index.
Identifiers: ISBN 9781532188763 (lib. bdg.) | ISBN 9781532189258 (ebook) | ISBN 9781098200237 (Read-to-Me ebook)
Subjects: LCSH: Zoos--Juvenile literature. | Zoo animals--Juvenile literature. | Zoology--Juvenile literature. | School field trips--Juvenile literature.
Classification: DDC 371.384--dc23

Table of Contents

Zo0 4

More Animals
at the Zoo 22

Glossary 23

Index 24

Abdo Kids Code 24

Zoo

It's field trip day. The class is going to the zoo.

The kids will see animals. They will learn about them too!

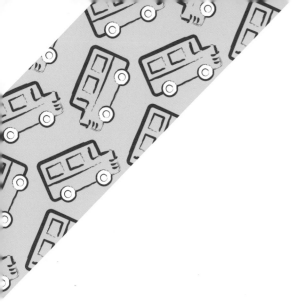
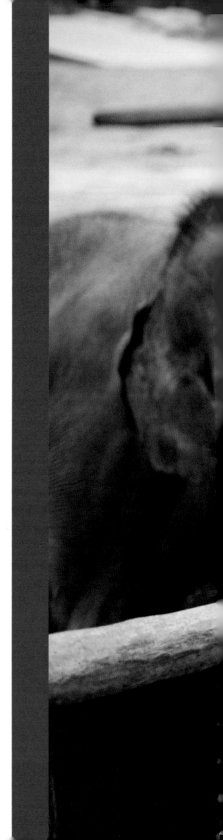

Rosa sees an elephant. It weighs 8,000 pounds (3,630 kg)!

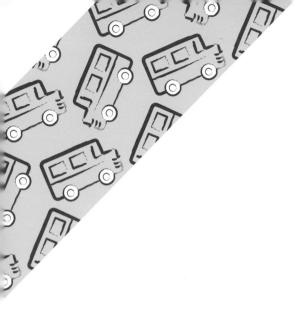

Kayla feeds the giraffe.

It is 18 feet (5.5 m) tall!

Owen sees the polar bear.

It is swimming.

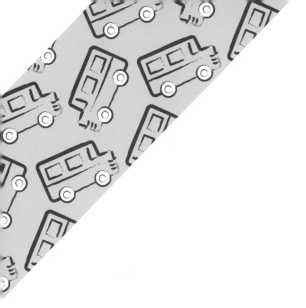

Lia holds a young tortoise. It will grow to be 200 pounds (90 kg)!

Zane meets a lemur. It holds his hand.

Lucy sees the lions. They have a loud roar!

Have you been to a zoo?

More Animals at the Zoo

brown bear

rhinoceros

shark

tiger

Glossary

roar
a loud, deep noise made by a lion.

tortoise
a turtle that lives on land and eats plants.

Index

animals 6, 8, 10, 12, 14, 16, 18

elephant 8

giraffe 10

lemur 16

lion 18

polar bear 12

tortoise 14

Visit **abdokids.com** to access crafts, games, videos, and more!

Use Abdo Kids code **FZK8763** or scan this QR code!